My Favorite Animal

I LOVE HORSES!

I0794158

By Beth Gottlieb

Gareth Stevens
PUBLISHING

Please visit our website, www.garethstevens.com. For a free color catalog of all our high-quality books, call toll free 1-800-542-2595 or fax 1-877-542-2596.

Library of Congress Cataloging-in-Publication Data
Names: Gottlieb, Beth, author.
Title: I love horses! / Beth Gottlieb.
Description: Buffalo, New York : Gareth Stevens Publishing, [2023] | Series: My favorite animal | Includes index.
Identifiers: LCCN 2022014041 (print) | LCCN 2022014042 (ebook) | ISBN 9781538283233 (paperback) | ISBN 9781538283257 (library binding) | ISBN 9781538283264 (ebook)
Subjects: LCSH: Horses–Juvenile literature.
Classification: LCC QL737.U62 G68 2023 (print) | LCC QL737.U62 (ebook) | DDC 599.665/5–dc23/eng/20220523
LC record available at https://lccn.loc.gov/2022014041
LC ebook record available at https://lccn.loc.gov/2022014042

First Edition

Published in 2023 by
Gareth Stevens Publishing
2544 Clinton St,
Buffalo, NY 14224

Editor: Kristen Nelson
Designer: Rachel Rising

Photo credits: Cover, pp. 1, 5, 11, 23, 24 Callipso88/Shutterstock.com; p. 7 WiP-Studio/Shutterstock.com; p. 9 Kwadrat/Shutterstock.com; pp. 13 Alyson W. Kast/Shutterstock.com; p. 15 Anastasija Popova/ Shutterstock.com; p. 17 Christy berry/Shutterstock.com; p. 19, 24 nazarovsergey/Shutterstock.com; p. 21 Victoria Shapiro/Shutterstock.com.

Printed in the United States of America

CPSIA compliance information: Batch #CWGS23: For further information contact Gareth Stevens, New York, New York at 1-800-542-2595.

Contents

Horses are my favorite animal.

They are beautiful.
They are covered in hair.

Horses can be white, black, or gray. They can be red or brown. They can have spots!

Horses have long hair on their head and neck. This is their mane.

Horses have big teeth.
They chew grasses.

Horses on farms eat oats and barley too.

Wild horses live in groups.
These are herds.

People ride horses.
They use special seats.
These are saddles.

I ride a horse.
I learn to trot!

I love horses!

Words to Know

mane

saddle

Index